FATIMA LAWRENCE

Playing the Lottery

Learn how it works and unlock secrets of improving your odds... because winning is way more fun!

Copyright © 2024 by Fatima Lawrence

All rights reserved. No part of this publication may be reproduced, stored or transmitted in any form or by any means, electronic, mechanical, photocopying, recording, scanning, or otherwise without written permission from the publisher. It is illegal to copy this book, post it to a website, or distribute it by any other means without permission.

Designations used by companies to distinguish their products are often claimed as trademarks. All brand names and product names used in this book and on its cover are trade names, service marks, trademarks and registered trademarks of their respective owners. The publishers and the book are not associated with any product or vendor mentioned in this book. None of the companies referenced within the book have endorsed the book.

First edition

This book was professionally typeset on Reedsy.
Find out more at reedsy.com

Contents

1 Introduction   1
2 Lottery Commissions   5
3 Ways to Play   8
4 Draw Game Odds   13
5 Instant Game Odds   20
6 Claiming Prizes   27

1

Introduction

Growing up my grandmother was a hairdresser, and I'd find myself going to the beauty shop after school to do homework so I could have a chance to earn a couple bucks by sweeping, laundering and folding towels, washing windows, or whatever else came up. You may feel like you tell your hairdresser things that you'd never tell anybody, and I'm here to tell you that you're not the only person who's doing so. As a quiet and incredibly nosy kid I heard all kinds of things, most of which was not appropriate for young ears, but the thing that always fascinated me the most was stories of lottery winners. In my suburban town along the rust belt, there were what felt like tons of people winning pretty good sums by playing the lotto. (None of what I overheard ever led me to believe that someone in our family won anything worth talking about… but, hey, you never know!) If you would ask me the best stocking stuffers or extra prizes in an Easter basket, I'd tell you that an instant lottery ticket was it!

For the majority of this book, we will talk about some of the math and statistics behind the lottery and how you can apply these methods to increase your chances of winning. I do not promise any guarantees of success or any secret formula that's going to "crack the code" of winning the lottery. I do intend to give you some new considerations and an understanding of how you could think differently about playing the lottery, hopefully with some greater outcomes your way!

Gambling is really no different from anything that gives us dopamine hits, which gives it the potential to turn from something fun into something addictive. Instagram can be fun and "checking my feed really quick" can easily turn into hours of scrolling leaving you what happened to the time. The consequences of excessive gambling can be a little more severe than an evening wasted on the couch. Gambling is a game and you should always remember to play responsibly. That said, recognizing gambling addiction can be challenging, but there are several telltale signs that might indicate a problem:

- Preoccupation with Gambling: Constantly thinking about gambling or planning the next gambling activity.
- Increasing Bets: Needing to gamble with larger amounts of money to achieve the same level of excitement.
- Chasing Losses: Trying to win back money that has been lost, often leading to further losses.
- Neglecting Responsibilities: Ignoring work, family, or personal responsibilities due to gambling.
- Secrecy: Hiding gambling habits or lying about the amount of time and money spent on gambling.
- Financial Problems: Experiencing financial difficulties due to gambling, including borrowing money or selling assets.
- Emotional Distress: Feeling anxious, depressed, or irritable when not gambling or when facing gambling losses.

INTRODUCTION

- Failed Attempts to Stop: Trying to cut back or stop gambling without success despite knowing the negative impacts.
- Gambling to Escape: Using gambling as a way to escape from stress, anxiety, or other problems.
- Loss of Control: Being unable to control the urge to gamble or feeling compelled to gamble despite negative consequences.

If these signs are present, seeking professional help from a counselor or support group specializing in gambling addiction can be an important step toward recovery.

Many states' "Bets Off" programs offer a strategy or tool used to help individuals manage their gambling behavior. This can include:

- **Self-Exclusion**: Programs or options that allow individuals to voluntarily exclude themselves from gambling venues or online gambling sites for a specified period, helping them take a break from gambling.
- **Gambling Blockers**: Software or tools that block access to gambling websites or limit the amount of time spent on gambling activities.
- **Support Resources**: Access to counseling, therapy, and support groups designed to help individuals address and manage their gambling addiction.
- **Account Limits**: Setting limits on gambling activity, such as deposit limits or spending caps, to control the amount of money and time spent on gambling.

The goal of these measures is to provide individuals with the tools and support needed to prevent or reduce gambling behavior and to promote healthier, more controlled gambling habits. You may also find helpful resources in your state that encourage healthy and responsible gambling.

So with the intention of having some fun and having the right mindset of where the lottery ought to be in your life, a fun game to take a shot at every so often, let's dive in!

2

Lottery Commissions

So when you're buying something it always begs the question, who is selling? Your state lottery commission has the right to regulate and oversee lottery games including monitoring game operations, setting rules and regulations, and ensuring compliance with legal standards. So as you read throughout this book, take the contents as general themes or principles, because the rules in your state may differ. The Lotto Commission has the duty to protect consumers by regulating retail locations, communicating to the public game details and results, and facilitating the process to claim prizes. Most state lotto programs exist to provide funding for education or infrastructure; however, each state does have the ability to decide the cause that will be supported by its funding. (As an example, for every dollar spent on tickets $.50 will go to prize pools, and the remainder will be used to support the department of education and the administrative costs associated with operating the lotto commission). Overall, the state lottery commission aims to operate lottery games in a way that maximizes benefits to the public while maintaining fairness and integrity.

Because Lottery Commissions are initially created to support a cause that benefits the public of that state such as infrastructure, veterans affairs, education, state's general fund, health and human services, etc., there is an underlying incentive for increasing revenues by selling more lottery tickets! Most states allow for advertising as a way for the Lottery Commission to increase awareness for Jackpots (maybe on a billboard you pass on your commute every day or a small sign tracking Jackpots hanging on the window of gas stations or convenience stores). You may also see advertising attempts with the newest instant games hanging right near the cashier at the grocery store or a vending machine that dispenses lottery tickets as you leave a superstore. Whatever the method, the Lottery Commission will generate more dollars for their cause if they're able to attract more participants.

There are five states that do not sell lottery tickets

- Alabama
- Alaska
- Hawaii
- Nevada
- Utah

Lottery Commissions also provide incentives for retailers who participate by selling lottery tickets. In general, retailers will receive a small commission of total ticket sales. Typically commissions are around 5% - this will vary by state, by the method of purchase such as a self-service vending machine or an employee operated terminal, product type could also impact commissions. In addition, retailers who sell winning tickets will also receive a percentage of winnings typically around 1% (subject to a maximum amount). Retailers could also be eligible for incentives when cashing out winning tickets.

In summary, Lottery Commissions exist to sell and administer lottery games while using the proceeds from the sale of lottery tickets to benefit the public. They incentivize retailers to ensure participants have access to purchase tickets. So now that we've covered the why, let's take a look at how.

3

Ways to Play

There are two ways to play the lottery 1) Draw Games and 2) Instant Games. Let's start with draw games because those are most often associated with "buying a lottery ticket." Draw games require a wager in the form of a set of numbers from a predefined pool and winning numbers are randomly drawn at a predetermined time. Draw games include Powerball, MegaMillions, and Cash4Life which are nationwide games administered by the Multi-State Lottery Association. Multi-state games are exactly as the name suggests, ticket sales are collected across multiple states to broaden the base of ticket sales and achieve larger jackpots. There are likely other games available only in your state such as Lotto, Pick 3, or Pick 4 games (branding will vary from state to state). However, because these games are only available for sale in a particular state, then the jackpots will not be nearly as large as those available in a multi-state game.

As mentioned earlier, your state lottery commission has an obligation to oversee the administration of lottery games while ensuring transparency and fairness. So each draw game will undergo the following process for each drawing and for each game:

- **Ticket Sales**: Players purchase tickets from authorized retailers or online platforms, choosing numbers based on the game's rules. Tickets are usually available for a specified period before the draw.
- **Ticket Validation**: Each ticket is validated to confirm it is a legitimate entry for the upcoming draw. This includes checking that the ticket is correctly formatted and paid for.
- **Draw Preparation**: Before the draw, the lottery commission prepares the drawing machines and ensures all equipment is functioning properly. This typically involves rigorous testing and verification procedures.
- **Draw Event**: At the scheduled time, a draw is conducted. This is often a public event to ensure transparency. Draw machines, which may use mechanical or electronic methods, randomly select numbers. The draw is usually witnessed by independent auditors or observers to ensure fairness.
- **Number Selection**: The drawing machine selects a set of winning numbers. For example, in a typical 6/49 game, six numbers are drawn from a pool of 49.
- **Results Announcement**: The drawn numbers are announced and published through various channels, such as the lottery's website, television, and other media. The results are also often posted at retail locations.

So let's say you decide you want to buy a Powerball ticket as you're leaving the grocery store one evening - who doesn't love "The Great American Daydream ®" right? You walk up to the customer service counter and say "I'll take $10 on a Powerball ticket." The assumption is that you want the system to generate a selection of numbers for you at random and you will be included in the next available drawing. Powerball drawings are three times per week on Mondays, Wednesdays, and Saturdays at 10:59 pm Eastern. There is a blackout period beginning 1 hour before the drawing where tickets are not available for sale (tickets are also not available for sale daily between 2am - 5am). You realize it's Thursday and the next drawing isn't until Saturday evening, so you're on the edge of your seat for the drawing two days from now. The multi-state lottery commission goes through the process and releases the winning numbers that have been chosen at random by a popcorn looking machine that throws a bunch of ping pong balls around. The winning numbers are announced and with great anticipation you check your tickets for a match.

I've used the Powerball as an example here but the process is the same for Mega Millions, Cash4Life, Pick 3, Pick 4, and other state sponsored lottery games. The difference between games will vary based largely upon factors including but not limited to:

- Multi State lottery games versus single state lottery games
- Number of drawings per week (some games might have multiple drawings per day)
- Number of matched numbers to win a prize
- Size of the prize tiers and Jackpot
- Number of years over which prize is paid (if applicable)
- Cost per play
- Overall odds
- Specific odds for winning at various prize levels

Now that we've provided an overview of draw games, let's turn our focus to the other type of game, instant games which can often be referred to as "Scratch-off" games.

Unlike draw games (which are the same games with the same odds at the same intervals), each instant game will follow a "lifecycle." The multi-state lottery commission does not sponsor instant games; the instant games you will find are specific to your state.

- **Concept and Design**: Lottery organizations or game developers design the game, including the theme, prize structure, and ticket design. This stage includes determining the odds of winning and the prize distribution.
- **Approval and Production**: The game design is submitted for approval to regulatory authorities to ensure compliance with legal requirements. Once approved, the tickets are produced, which involves printing and packaging.
- **Distribution**: The tickets are distributed to retailers and sold to the public.
- **Sales and Gameplay**: Players purchase tickets and scratch or reveal numbers to determine if they've won. This phase continues for the duration of the game's active period.
- **Prize Claims**: Winners claim their prizes according to the game's rules. Prizes can be claimed at retail locations, regional offices, or by mail, depending on the amount and type of prize.
- **End of Game**: The game concludes when all tickets are sold or the game's end date is reached. The lottery organization may then conduct audits to ensure all prizes have been awarded correctly.

So the key difference here is that an instant game is printed and then distributed. Let's imagine the lottery commission decides to print a game with 2 million tickets for a particular instant game. Of the 2 million tickets printed there are 500,000 winners. That's it for the particular game from the time the game is released until the time the lottery commission announces "end of game."

We will cover this more in the chapter covering Instant Game Odds, but the overall odds are always the overall odds. In this particular game, participants have a 1 in 4 chance throughout the entirety of the game. However, as the number of prizes changes and the number of remaining tickets continues to decline, the specific odds of winning a particular prize may improve.

4

Draw Game Odds

I'll start with a bit of a warning here - as the title of this chapter suggests we are going to have to talk about statistics. Maybe this is the type of reading that will put you to sleep or maybe you can never get enough of numbers - either way, let's jump in!

Let's take a step back and consider what it actually means (and the odds of) winning the jackpot for either a Powerball ticket or Mega Millions ticket. Both games require that you match 5 numbers AND match the Powerball or Mega Ball. So if you think about it, you really need to win the lottery "twice" in order to win the jackpot amount.

Statistically, this is referred to as solving for combinations, where there are 5 combinations of numbers that are derived from 69 variables (for Powerball). There are 11,238,513 ways to choose 5 numbers from 69. Pretty impressive to hit all five of these; however, you still need to get the Powerball in addition to all five numbers to win. Since there are 26 possible choices for the Powerball, multiply the possible combinations of 5 numbers by 26:

PLAYING THE LOTTERY

Total possible Powerball Jackpot tickets = 11,238,513 × 26 = 292,201,338

So what does that mean? For $2 you have an approximate 1 in 292.2 million chance of hitting the Jackpot.

The odds for Mega Millions are a little bit different because there are more possible combinations for the 5 numbers because Mega Millions uses 70 numbers instead of 69. There are 12,103,014 ways to choose 5 numbers from 70. Slightly different from Powerball, the MegaMillions uses 1-25 for the Mega ball instead of 26.

Total possible MegaMillions tickets = 12,103,014 × 25 = 302,575,350

So what does that mean? For $2 you have an approximate 1 in 302.6 million chance of hitting the Jackpot.

Those odds are incomprehensible for the human mind because the chances are so small. The incredibly unfavorable odds of winning a multi-state Jackpot still lure folks into buying tickets because the Jackpot amounts are so huge. At some point, somebody wins the Jackpot, but the realistic probability is that it's not going to be you.

I'm not a medical professional or a mental health professional of any kind, but let's look at this upside down. Humans hold on to hope that there is a possibility of winning the Powerball or Mega Millions Jackpot because the outcomes are perceived to be favorable - you want, wish, pray, desire, would do absolutely anything for it to happen to you! But let's think about something perceived as adverse that you would not want to happen to you that has similar (or even more likely odds). Consider an expectant mother who is concerned that her child could have a rare genetic disorder; she has extensive conversation with her

DRAW GAME ODDS

OB/GYN who tells her that there are 1 in 1,000,000 deliveries that report this particular diagnosis each year. I don't know about you, but I'd leave that appointment with a sigh of relief! Odds are your child is not going to be born with that particular genetic disorder. If you assume that there are approximately 300 million people in the United States of America, then there are probably no more than 300 people in the entire country that have that particular diagnosis. Now consider that against the odds of winning a Powerball or Mega Millions Jackpot, approximately 1 in 300,000,000.

There is always somebody that eventually wins that giant jackpot though... I digress! Let's move on.

So how did I hear all these stories of lottery winnings growing up? In-state draw games! Let's pick a couple examples of what a state only game might look like. Keeping in mind that each Lottery Commission, as we mentioned earlier, is going to have their own discretion for creating and administering games that will attract participants and fund the underlying cause funded by Lottery proceeds. Games will vary from state to state.

Let's assume a statewide "Draw Game #1" where there are 6 numbers ranging from 1-44. Matching all 6 numbers wins you the jackpot! In this instance, a $1 ticket will buy two sets of numbers or "plays." Let's assume that this particular game has drawings twice per week, and the Jackpot will most likely range from $1,000,000 to $10,000,000.

Total possible lotto tickets for "Draw Game #1" = 7,059,052

Because each $1 play means two sets of 6 numbers, then the total possible of 7,059,052 is divided by 2.

So what does that mean? For $1 you have an approximate 1 in 3,529,526 chance of hitting the Jackpot.

One last example so we can show the differences between two types of in-state games. Let's assume there is another game "Draw Game #2" where there are 5 numbers ranging from 1-39. Matching all 5 numbers wins you the jackpot! In this instance, a $1 ticket will buy you one set of numbers or "plays." This particular game has drawings daily, and the Jackpot will most likely range from $100,000 to $1,00,000.

Total possible lotto tickets for "Draw Game #2" = 575,757

So what does that mean? For $1 you have an approximate 1 in 575,757 chance of hitting the Jackpot. Notice the substantial change in odds of winning the Jackpot from reducing the population of numbers from 1-44 down to 1-39 and the number of winning numbers from 6 down to 5.

So far we have only been discussing the idea of what it would take for a participant in a draw game to hit the jackpot. There are, however, many other ways to win smaller prizes when playing a lottery game.

DRAW GAME ODDS

Powerball Odds		
MATCH	PRIZE	CHANCES 1 IN
5 white balls + Powerball	Jackpot	292,201,338
5 white balls	$1,000,000	11,688,054
4 white balls + Powerball	$50,000	913,129
4 white balls	$100	36,525
3 white balls + Powerball	$100	14,494
3 white balls	$7	580
2 white balls + Powerball	$7	701
1 white balls + Powerball	$4	92
0 white balls + Powerball	$4	38

Overall Powerball Odds are 1 in 25

Mega Millions Odds		
MATCH	PRIZE	CHANCES 1 IN
5 white balls + Mega Ball	Jackpot	302,575,350
5 white balls	$1,000,000	12,607,306
4 white balls + Mega Ball	$10,000	931,001
4 white balls	$500	38,792
3 white balls + Mega Ball	$200	14,547
3 white balls	$10	606
2 white balls + Mega Ball	$10	693
1 white balls + Mega Ball	$4	89
0 white balls + Mega Ball	$2	37

Overall Mega Millions Odds are 1 in 24

State Lotto Game #1 Example Odds		
MATCH	PRIZE	CHANCES 1 IN
6 of 6 numbers	Jackpot	3,529,526
5 of 6 numbers	$715 (average amount)	15,480
4 of 6 numbers	$29 (average amount)	335
3 of 6 numbers	Free Ticket (QP)	21

Overall State Lotto Game #1 Odds are 1 in 20

State Lotto Game #2 Chances		
MATCH	PRIZE	CHANCES 1 IN
5 of 5 numbers	Jackpot - minimum of $50,000	575,757
4 of 5 numbers	$250	3,387
3 of 5 numbers	$10	103
2 of 5 numbers	$1	10

Overall State Lotto Game #2 Odds are 1 in 9

So what is the difference between overall odds and odds? Overall odds tells you what are the chances you will achieve at least some type of "winning ticket." In most cases the overall odds are going to match (more or less) the odds of the lowest prize in the draw game. It's calculated as adding all of the odds for each prize together to get at the "overall odds." Note that in most cases, the overall odds will most likely equate to a participant "winning their money back."

So what does it all mean? When playing a draw game, the odds are the odds. There is a period of time where tickets are sold (in general, not all possible combinations will exist in the total population of tickets sold for a particular drawing). Someone either has the winning ticket or they don't. The only way for you to improve your chances… is to buy more tickets.

5

Instant Game Odds

Same as the last chapter - you know the drill. Time to buckle up for some math!

So like draw games, instant games have overall odds that are determined at the creation of the game. Simply put: there are a finite number of tickets printed and there are a finite number of winning tickets printed. However, as the number of tickets remaining will change, your odds of winning top prizes will change over time... sometimes favorably... and sometimes unfavorably.

While not a perfect analogy, the famous Hasbro board game of Battleship© is a good way for us to demonstrate odds for instant games. If you're not familiar with the game, each player has a naval fleet of five ships hidden on their own grid with an X-axis of numbers ranging from 1-10 and a Y-axis of letters ranging from A-J. The object of the game is to guess the location of the other player's ships and sink them before he/she is able to sink all of your ships. I know there are some of you getting excited and want to go play a round right now, but hang tight! The point here is there are 100 places that a ship can go - so your overall odds of striking your opponent's target are 1:100... no matter what. The size of the "ocean" is the same whether you've made zero guesses or thirty guesses.

Sample Battleship© Board:

	1	2	3	4	5	6	7	8	9	10
A										
B										
C										
D										
E										
F										
G										
H										
I										
J										

Again, overall odds will always remain overall odds. There's always going to be the same number of winning tickets for a particular game. Something to remember is that tickets are printed at random. If we assume overall odds of winning were 1 : 3, then if I buy three tickets I should have at least one winner in there, right? Not necessarily because tickets are printed at random. The only way this would work out is if you were to purchase every ticket issued for a particular game, then you'd be able to say that 1.3 tickets are winners. Since that's not going to necessarily happen, then those are overall odds of winning because

1:3 tickets printed will be a winner.

So let's use an example to illustrate how this could play out. There are two games that we can compare side by side one started in the spring of last year and is named "Pot of Gold" featuring a leprechaun at the end of a rainbow advertising top prizes of up to $2,000,000! The game has been going on for about 18 months now and tickets have been selling consistently throughout this time.

Now the lottery commission has a new game out just in time to celebrate Thanksgiving with "Winnings to be Thankful For" and also advertises top prizes of $2,000,000.

Both games have the exact same odds: 3.04.

So now we have to determine the number of prizes in the game. Lottery websites will post some type of grid to show the distribution of prizes. This will allow us to calculate the total number of tickets.

Total tickets = (Total winning tickets) x (Overall Odds)

3.04 x 1,076,295 = 3,271,937

Each state will publish this information in different combinations, but if you have two of the numbers then you can solve for the third.

Now a side by side comparison of each of the two games we mentioned earlier.

Pot of Gold

INSTANT GAME ODDS

Prize	Total Prizes	Unclaimed Prize
$20	463,709	128,309
$40	327,303	88,446
$50	140,485	37,410
$100	136,387	36,138
$200	5,451	1,472
$500	1,801	478
$1,000	1,112	297
$5,000	25	8
$10,000	11	3
$20,000	5	0
$100,000	4	3
$2,000,000	2	1
Total Prizes	**1,076,295**	**292,565**

Overall Odds: 3.04

Total Tickets: 3,271,937

Winnings to be Thankful For

Prize	Total Prizes	Unclaimed Prize
$20	463,709	463,709
$40	327,303	327,303
$50	140,485	140,485
$100	136,387	136,387
$200	5,451	5,451
$500	1,801	1,801
$1,000	1,112	1,112
$5,000	25	25
$10,000	11	11
$20,000	5	5
$100,000	4	4
$2,000,000	2	2
Total Prizes	1,076,295	1,076,295

Overall Odds: 3.04

Total Tickets: 3,271,937

So at first glance, looking at these two games most folks are going to gravitate towards the $2,000,000 prize at the bottom and see how many are left. Since "Winnings to be Thankful For" has two of the top prizes remaining, then that must be the one that I want to buy, right?

Not necessarily - let's do some math to see the why behind it. The next important concept that we will introduce here is an estimate on the number of tickets remaining.

The estimated number of tickets remaining for "Winnings to be

Thankful For" is easy… all of them! There are approximately 100% tickets outstanding because it's a brand new game.

Let's pause and go back to our Battleship© example. Suppose you're at a family gathering with all of your cousins, and there's two games of Battleship© going on. Two opponents are tired of playing and want you to take over for them. One game has just started and there have been zero guesses. The second game has been going on for a little bit, and there have been 25 guesses (let's assume that to keep things even, neither opponent has gotten a "hit" out of these guesses) Again, not the perfect analogy, but you probably want to take over the 2nd game because you still have a 1:100 chance of hitting your opponent; however, only 75 potential targets remain. OK, back to estimated tickets remaining:

Estimated Tickets Remaining = Unclaimed Prizes / Total Prizes

If we do the same math for "Pot of Gold" then the number of remaining tickets is much less than 100% because the game has been selling tickets for about a year and a half now.

Pot of Gold Estimated Tickets Remaining

292,565 / 1,076,295 = 27.2%

So what does that mean? There are 2 jackpot prizes available for "Winnings to be Thankful For" and 1 jackpot prize available for "Pot of Gold." The overall odds of buying a winning ticket for each game is 1 in 3.04.

"Pot of Gold" is a better buy because over 75% of the tickets have already been bought!

PLAYING THE LOTTERY

If you were to play the "Winnings to be Thankful For" game, then your odds of winning the Jackpot are:

2 in 3,271,937

If you were to play the "Pot of Gold" game, then your odds of winning the Jackpot are:

1 in 889,398

Note that the 889,398 here comes from multiplying the estimated percentage of tickets remaining times the total number of tickets printed.

Now, this math will ring true for each of the individual prize levels.

Another way to easily think about the concept here (if you're not able to have a spreadsheet in front of you as you're thinking about what type of instant lottery ticket to buy). Ask yourself the question(s):

- What is the 1st date tickets were available for sale?
- How many of the remaining tickets are outstanding for the 3 largest prizes?
- Are there a lot of "unclaimed prizes" remaining?

These questions could be a good way for you to think about which ticket could be best as you're considering the right instant lottery ticket to buy.

6

Claiming Prizes

Understanding your odds doesn't mean there are any guarantees of winning, but there are some chances that could be better than others. I'll assume that if you do win some meaningful prize money, you'd want to do the most prudent thing with your winnings.

If claiming large enough prizes for draw games or instant games, depending on the size of the prize and the rules of the game, there may be an option for a cash payout (less than the total prize amount) or a series of payments over a period of years (20, 25, 30 years).

So let's use an example of a $2,000,000 prize payable in one of two ways:

1. $100,000 / year for 20 years
2. $1,200,000 lump sum payout

When making this decision consider a few factors:

1. Human behavior is not always rational (especially spending behavior)
2. Investment returns
3. Life expectancy

Most people would look at those choices and assume the best option is to take a lump sum, no matter what. You're immediately losing 40% right off the bat, then you're going to lose another (estimated) 40% of that to taxes. So now we've taken what could have been $2,000,000 and turned it into $720,000. I understand that if $720,000 fell out of the sky... my first thought certainly wouldn't be "Aw, shucks, I can't believe this isn't $2,000,000!"

Most finance professionals would tell you that you'd wind up with more money if you took a lump sum and invested it.

Let's be realistic here and assume that you spend $220,000 to pay off your house, buy a nice car, and go on a nice vacation. You've got $500,000 left.

If you invested $500,000 for 20 years it would take an annual rate of return of 7.2% (after tax) it would equal the $2,000,000 you originally won.

Meanwhile lets not forget you have to look at an account balance for 20 years and not spend any of it. (In my opinion, that's the important part people don't consider).

Conversely, you could take the annuity option, You receive $100,000 / year for 20 years.

You set up a self-directed brokerage account and pick a few ETFs and invest $2,000 per month for the next 20 years. In this instance, let's remember your tax liability on $2,000,000 over 20 years is a better deal than your tax liability on $1,200,000 in one year. Let's assume you lose 26% per year to tax and 24% per year goes to investing. You've got an extra $50,000 per year to spend freely however you'd like.

Now if we assume the same 7.2% annual rate of return (after-tax) as the other example, then $2,000 per month invested for 20 years equals just over $1,000,000. Personally I'd rather get a "$50,000 bonus" every year for 20 years, live my life, help my kids or my community, and still have $1,000,000 at the end. I also think there's less temptation and pressure when receiving payments over time than having to be a steward of assets.

Depending on your circumstances, you may still choose the lump sum, you may find that an annuity option is better for you.

Lastly, read the fine print on deadlines for claiming prizes because most draw games require you to claim your prize within six months. If you want the cash payout (at the reduced rate) then you may have to claim within two months rather than six months.

If you win an instant game, then you have six months from the "end date" of the game. Once the lottery commission decides that a particular game has run its course, then they reserve the right to stop the distribution of the tickets and terminate the game. Once they announce the end of the game, then your clock is ticking to claim. Not necessarily from the date you bought the ticket and won.

If you do win a life changing amount of money, it is in your best interest to seek the advice of a tax professional, an investment professional, and an attorney. Listen to their advice, but also remember that they work for you, and it's OK to challenge their recommendations if you don't like how something sits with you.

I do hope this gives you some additional things to think about or at the very least some different ways of thinking about them. Best of luck to you and play responsibly!

www.ingramcontent.com/pod-product-compliance
Lightning Source LLC
Chambersburg PA
CBHW070957220526
45471CB00007B/3065